Work 247

小僵尸
Tiny Zombies

Gunter Pauli

[比] 冈特·鲍利 著

[哥伦] 凯瑟琳娜·巴赫 绘

朱 溪 译

上海远东出版社

丛书编委会

主　任：贾　峰

副主任：何家振　闫世东　郑立明

委　员：李原原　祝真旭　牛玲娟　梁雅丽　任泽林

　　　　王　岢　陈　卫　郑循如　吴建民　彭　勇

　　　　王梦雨　戴　虹　靳增江　孟　蝶　崔晓晓

目录

Contents

一只鲍鱼坐在海底，静静地打造坚硬且有光泽的贝壳，这是她的家。一个病毒在一旁观察并评论道：

　　"祝贺你拥有这样的家！它那么轻巧又那么坚固，既干净又有光泽，实在让我印象深刻。"

An abalone is sitting on the ocean floor, quietly working away at making the hard, shiny shell that is her home. A virus is watching and comments,
"Congratulations on your home. I'm impressed: it is so light and yet sturdy. And so clean and shiny."

一只鲍鱼坐在海底……

An abalone is sitting on the ocean floor ...

你的壳也不会破碎

your shell won't break

"嗯，我们几千代人花了几百万年的时间才破解了建造房屋的艺术，那就是怎么充分利用海洋里到处漂浮的这些小砖块。"

"即便用锤子砸，你的壳也不会破碎，这是真的吗？"

"是的。我们所做的不是对抗打击，而是滑动并伸展我们的小瓷砖来吸收这种巨大的力量。"

"Well, it took us thousands of generations and millions of years to crack the art of home building with these tiny bricks that are floating everywhere in the sea."

"Is it true that when hit with a hammer, your shell won't break?"

"Yes. What we do, instead of resisting the blow, is to slide and stretch our tiny tiles, to absorb such a great force."

"这太明智了：顺势化解力量，而不是抵抗它。"

"是的，这看起来也正是我安排生活的方式——身边有那么多像你一样的病毒围绕着……"

"好吧，没有多少人喜欢我们病毒是地球上最丰富的生命形式这一现实。当地球上有生命存在时，我们就已经存在了。人们宁愿无视我们，甚至试图杀死我们，也不愿面对我们。"

"That is wise: go with the force, do not resist the force."
"Yes, and it seems to also be the way I need to organise my life – with so many viruses like you around…"
"Well, not many people like the idea that we are the most abundant form of life on Earth. We have been around for as long as there has been life on Earth. People would rather ignore us, or try to kill us than face us."

地球上最丰富的生命形式……

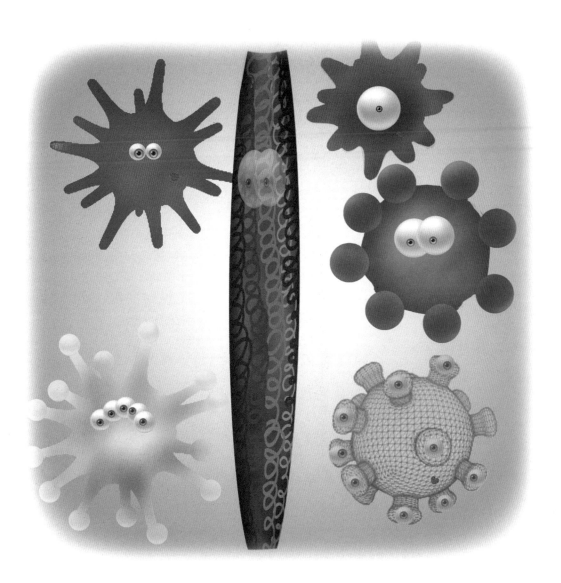

Most abundant form of life on Earth ...

我们不能生宝宝……

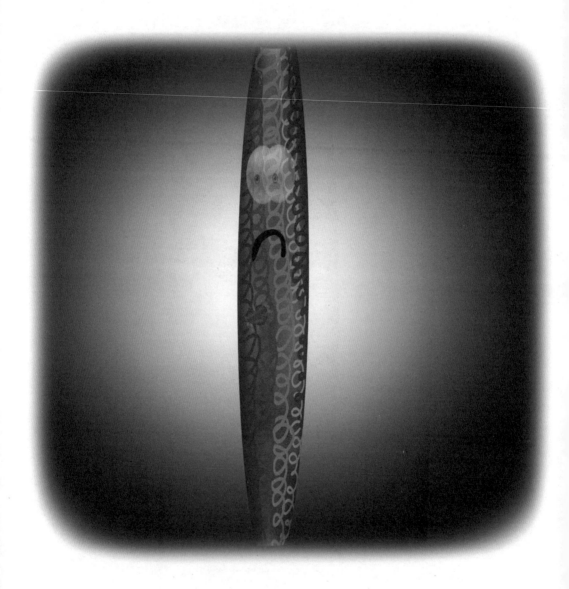

We can't have babies ...

"可是你甚至都不是活体，而且无法独自生存。"

"的确如此，但我们也的确会四处移动，养活自己。只是我们不能生宝宝……"

"听说人类可以将你们驯化，就像他们对小狗那样？"

"But you are not even a living organism, and cannot survive on your own."

"True, but we do move around and feed ourselves. It is just that we can't have babies …"

"And is it true that people can teach you tricks, like they do with dogs?"

"你问的可真巧！请放心，很快将不再有任何演出了。人类知道我们中的一部分很擅长……"

"……让他们生病，的确是的，你们会导致很多疾病。"

"你为什么只听坏消息？试着改变吧，去听听好消息，可以吗？"

"Funny that you should ask! Rest assured, there won't be a show anytime soon. People know that some of us are very good at … "

"… Making them sick, yes. You cause many illnesses."

"Why do you always only listen to the bad news? Listen to the good news for a change, will you?"

你们会导致很多疾病……

You cause many illnesses ...

我们能制作电池……

We can make batteries …

"好消息？很难忘记你们能引起癌症、麻疹、登革热、流行性感冒……只要你说得出的。难道你希望我为此祝贺你吗？"

"可是你知道我们能制作电池吗？"病毒问。

"电池？我太震惊了！我们急需生物电池来替代那些用肮脏的化学品和大量金属制成的电池，它们制造了麻烦。"

"Good news? It is hard to forget that you can cause cancer, measles, dengue fever, influenza... you name it. Do you expect me to celebrate that?"

"But did you know that we can make batteries?" Virus asks.

"Batteries? I am surprised! Bio-batteries are badly needed to replace those made with dirty chemicals, and lots of metal, that cause a mess."

"我们现在就能制作电池。就像你一样，我们从水中组装微小的颗粒。当然，我们会部署一支训练有素的病毒军队。"

　　"训练有素？你们还能受训？我以为你们注定只会杀戮。"

"We can make batteries right here, right now. We assemble tiny particles from water, just like you do. Of course, we deploy a whole army of us well-trained viruses."

"Well-trained? You are able to be trained? I thought you are only destined to kill."

一支训练有素的病毒军队……

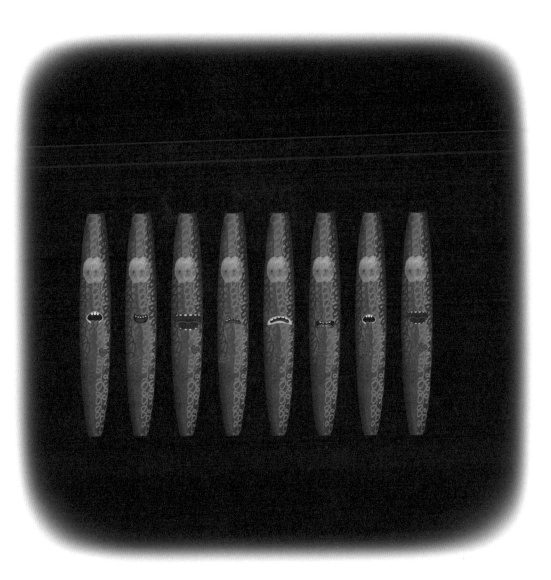

A whole army of us well-trained viruses ..

你们像贵宾犬一样被繁育吗？

Are you bred like poodles?

"那可一点也不好听！我理解许多人感到困惑和恐惧，因为要理解我们是什么、我们能做什么并不容易。但老实说，我们可以按照我们所学的去做。"

"你们像贵宾犬一样被繁育吗？"

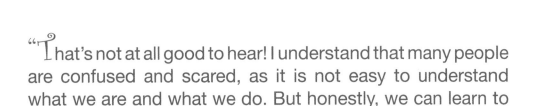

"That's not at all good to hear! I understand that many people are confused and scared, as it is not easy to understand what we are and what we do. But honestly, we can learn to do exactly as we are told."

"Are you bred like poodles?"

"你似乎真的不太喜欢我。但是不用担心；我知道你来自哪里。我们也是最近才被培育用来生产生物电池。"

"我对你们这些小僵尸印象太深了。下次我开电动车的时候，一定会去看看你们是不是在里面。"

……这仅仅是开始！……

"Υou really don't seem to like me much? But don't worry; I understand where you come from. We've only recently been bred to make bio-batteries."

"I am so impressed with you little zombies. Next time I drive an electric car, I'll check to see if you are inside it."

... AND IT HAS ONLY JUST BEGUN!...

······这仅仅是开始！······

... AND IT HAS ONLY JUST BEGUN! ...

磷是三磷酸腺苷（ATP）的活性成分。它也形成了DNA的基础，在细胞壁结构中很重要。没有磷，鲍鱼将无法生长出外壳。

Phosphorus is the active component of ATP; it also forms the backbone of DNA and is important in the structure of cell walls. Without phosphorus the abalone would not be able to produce its shell.

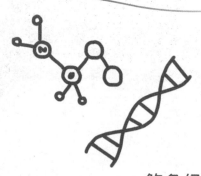

鲍鱼经过进化后，其DNA产生了一种蛋白质，可以从富含矿物质的水生环境中提取钙分子并将其贮存在体内的有序层中。

Abalone evolved so that its DNA produces proteins that extract calcium molecules from the mineral-rich aquatic environment and deposit it in ordered layers on its body.

It is estimated there are more than 10^{31} bacteriophages on the planet, more than every other organism on Earth, including bacteria. Up to 70% of marine bacteria may be infected by bacteriophages.

据估计，地球上有超过 10^{31} 个噬菌体，比地球上其他任何生物体（包括细菌）都多。多达 70% 的海洋细菌可能被噬菌体感染。

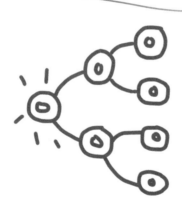

Viruses are acellular. They do not reproduce through cell division, but use a host cell to produce millions of copies of themselves. Biologists have not considered viruses as a form of life.

病毒是无细胞的。它们不通过细胞分裂繁殖，而是使用宿主细胞产生数百万个自身副本。生物学家尚未将病毒视为一种生命形式。

Viruses are a natural means of transferring genes between different species. This increases genetic diversity and promotes evolution. Viruses are one of the largest reservoirs of unexplored genetic diversity on Earth.

病毒是在不同物种之间转移基因的自然手段。这增加了遗传多样性并促进了进化。病毒是地球上最大的未开发的遗传多样性库存之一。

A pathogenic ganglioneuritis virus will kill all abalone on land-based abalone farms. In the wild, mortality rates of 90% will be achieved in just 2 weeks. Species that are immune then evolve and repopulate the reefs.

致病性的神经节神经炎病毒会杀死陆地鲍鱼养殖场中的所有鲍鱼。在野外，仅需2周，死亡率即可达到90%。具有免疫力的物种随后进化并重新居住在珊瑚礁中。

世界海洋中包含20万种病毒，比科学家先前记录的数量多了两个数量级。研究人员还在北冰洋发现了一个意想不到的病毒多样性区域。

The world's oceans harbour 200,000 virus species, two orders of magnitude more than scientists had previously recorded. Researchers also found an unexpected pocket of viral diversity in the Arctic Ocean.

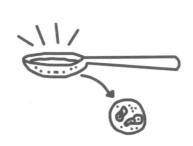

每勺海水中都充满了数百万个病毒。它们对人类无害，却可以感染鲸、甲壳类动物和细菌。同时，海洋病毒能使碳从海洋表面到达海洋深处。

Every spoonful of seawater is filled with millions of viruses. Harmless to people, they can infect whales, crustaceans and bacteria. At the same time, marine viruses drive carbon from the ocean's surface to its depths.

Are all viruses bad?

所有病毒都有害吗?

Are you scared of a virus?

你害怕病毒吗?

Are you only considered a form of life if you can make babies?

只要能繁衍后代就可以被视作为一种生命形式吗?

Are you prepared to listen to good and bad news?

你准备好听好消息和坏消息了吗?

Do It Yourself!

自己动手!

How do people react when you tell them you have a virus? People think you are sick. What will people think when you share the fact that you know how to train viruses to behave, and to even produce for you? We are dependent on viruses and bacteria, so how is it that these have such a bad image? Why do we only associate these with disease and death? Discuss these issues with people around you. Ask their opinions, and list all the fears they have. Next, compile a list to demonstrate what life would be like without viruses and share it.

当你告诉别人你感染了病毒,人们会如何反应? 人们会认为你病了。当你说你了解如何训练病毒,甚至能让它们为你所用时,人们会怎么想? 我们依赖病毒和细菌,可是病毒和细菌的形象为什么这么差? 为什么我们只将它们与疾病和死亡相联系? 与周围的人讨论这些问题。

听听他们的意见,并列出他们的担忧。列一份清单,说明没有病毒,生活又会怎样,并和别人分享。

学科知识
Academic Knowledge

生物学	鲍鱼属于海螺类，也是海洋腹足纲软体动物；病毒具有"病毒囊膜"，即保护膜；鲍鱼的肉就是腹足肌；病毒是地球上最丰富的生命形式；生殖作为生命分类的参数。
化 学	抗病毒和疫苗；珍珠质；电池的化学成分。
物 理	鲍鱼壳是几何、弹力和摩擦力相互作用的产物；病毒在空气中漂浮，在水中和皮肤表面生存；病毒几乎可附着在任何东西上，非常易于迅速传播；色彩斑斓极度闪耀的珍珠质；灰尘微粒对健康有害。
工程学	病毒用于制造疫苗、抗体和大多数转基因生物；具有黏着蛋白的微型瓷砖比陶瓷产品更坚固；海水中砖的液态组合；电池设计。
经济学	疫苗的盈利能力；病毒攻击及大流行给社会造成的代价；计算机病毒的成本，以及防止信息系统受到病毒攻击所需的投资；电池的回收成本，以及无法回收，从而造成威胁人类和生态系统健康的成本。
伦理学	游说者敦促决策者做出有利于他们的政治决策；社会如何决定抗击新冠病毒及"封城"，而登革热或HIV等其他流行病却不会导致同样的情况；间接伤害与意外后果；在南非非法捕捞鲍鱼导致该物种濒危；抵抗或分散力量；为了取悦人类而训练动物。
历 史	在南非布隆波斯洞穴中有100 000年历史的沉积物中发现了鲍鱼壳；20世纪50年代后期日本开始养殖鲍鱼；1974年发现了甲型肝炎病毒。
地 理	通过迅速在地球循环的急流，病毒从海洋传播到空气中；在冷水中发现的鲍鱼种类。
数 学	数学模型能使疫情可视化并预测疫情蔓延程度；鲍鱼具有断开式螺旋结构。
生活方式	需要接触以产生抗体；通常人们对坏消息而非好消息的兴趣和关注度更高。
社会学	暴发，流行和全球流行之间的区别；"病毒"一词来自拉丁语，意思是"毒液"，表明人们对它们有害的先入之见，尽管大多数病毒都是有益的；鲍鱼珠也已经被收集了数百年。
心理学	媒体灌输恐惧感，并通过反复沟通，引起行为改变；惊讶的影响；信息冲突导致混乱。
系统论	过度捕捞和偷猎减少了野生种群的数量，以至于现在人们消费的大部分鲍鱼是养殖的；鲍鱼以海藻为食，随着生态系统的丰富而自然生长，导致粉红鲷鱼、濑鱼和参孙鱼的数量增加。

情感智慧
Emotional Intelligence

病 毒

病毒欣赏鲍鱼的努力，并热情地给予祝贺。他问了一些问题，想知道鲍鱼壳是如何长成的。为了建立自信，他执着地展开对话。遇到恐惧时，他首先从力量的角度讲起，清楚地表明病毒在数量上超过了任何生命形式。同时，他也表明了自己的认知局限。这引发了鲍鱼的质疑，对此病毒冷静地接受。病毒直接问鲍鱼为什么只听坏消息。当鲍鱼坚持时，病毒彻底改变了话题。他开始谈论电池，并将组装过程与鲍鱼的工作进行比较。即使受到负面评论以及缺乏信任的双重轰炸，病毒表面上仍保持镇定。

鲍 鱼

尽管鲍鱼知道病毒有可能杀死她，但是当病毒祝贺她并且称赞她建造房屋的能力时，她还是与病毒进行闲聊。她花时间解释了自己极富韧性的建筑系统的原理。鲍鱼很快意识到自己正面对一种可能致命的病毒，于是开始担心。她直言不讳地质疑病毒是不是"活物"。随着对话的发展，她询问了更多有关病毒具备的学习人类所教内容的能力。随着病毒改变话题，鲍鱼不再想这些负面因素，而是跟着病毒的思路。最后，她甚至能够提出有趣的评论，这表明他们建立了更加信任和同情的关系。

艺术
The Arts

让我们画一些病毒。首先收集各种病毒的图像。注意它们鲜艳的色彩及不同的形态。病毒是三维的，因此不容易在平面纸上表现出来。你可以学一学如何创建三维透视图，来呈现你所收集的病毒图像。病毒还有显著的几何形状，这有助于你绘制。掌握了绘制病毒的技巧后，再添加一些颜色。

思维拓展
Systems: Making the Connections

　　大自然具有就地取材加以利用的卓越能力。鲍鱼会捕捉钙和磷的小分子，并在微米级的层面加以组装，使之在室温下形成结构。病毒采用完全相同的组装技术，可用于创建电池的生产系统。但我们似乎认为病毒只有一种著名特征，即引发感染，却忘记了病毒是地球上最古老的生命形式这一事实。我们贬低病毒，并把它们归为敌人，需要发起战争。这种观点以及我们向孩子灌输的恐惧，不仅没有道理，而且蒙蔽了我们的双眼。我们需要改变对待病毒的方式，并将其视为我们世界、日常生活以及生态系统的一部分，因为没有它们，我们将无法存活。大多数病毒是良性的，许多还是最有用和最被需要的，为什么我们还要消灭它们呢？我们面临的挑战是去教育人们重新认识病毒，因为目前只有少数人了解病毒的有益特性，以及我们在多大程度上需要病毒来维持生命和控制细菌。在扭转不可持续的生产和消费方面，只有在充分了解病毒并利用病毒来满足人类的需求之后，这些解决方案才变得切实可行。

动手能力
Capacity to Implement

　　不像细菌可以在人工培养基中生长，病毒需要活的宿主细胞才能进行复制。动物病毒的培养对于病原性病毒的鉴定与诊断以及疫苗的生产都很重要。你想学习如何培养病毒吗？这个过程需要成人监督，所以请让家长或老师帮你。在继续下一步之前，请先研究这个过程。你需要一些鸡蛋。将每个鸡蛋放在光线下，以便观察气囊的位置。用大头针在气囊区域内的蛋壳上打一个小孔。给每个鸡蛋接种病毒种株。用指甲油密封小孔，并在适当的温度下孵化鸡蛋。在鸡蛋旁边放一盘水，以防止鸡蛋破裂。该病毒会在鸡蛋膜的细胞中自我复制。

故事灵感来自
This Fable Is Inspired by

安吉拉·贝尔彻
Angela Belcher

安吉拉·贝尔彻就读于美国加利福尼亚大学圣巴巴拉分校，1991年获得创意研究学院学士学位，1997年获得化学博士学位。2006年她获得麦克阿瑟基金会奖学金，并被授予美国年度研究领导者称号。她一直担任生物工程系主任，现在还是詹姆斯·梅森·克拉夫茨讲席生物工程教授。贝尔彻博士研究自然过程，以用于能源、环境和医学的材料与设备设计。她使用定向进化技术来培育和组装可用于制造太阳能电池片、电池新材料以及医学诊断的新型材料。

图书在版编目(CIP)数据

冈特生态童书.第七辑:全36册:汉英对照 /
(比)冈特·鲍利著;(哥伦)凯瑟琳娜·巴赫绘;
何家振等译.—上海:上海远东出版社,2020
ISBN 978-7-5476-1671-0

Ⅰ.①冈… Ⅱ.①冈… ②凯… ③何… Ⅲ.①生态
环境 – 环境保护 – 儿童读物—汉英 Ⅳ.①X171.1-49

中国版本图书馆CIP数据核字(2020)第236911号

策　　划 张　蓉
责任编辑 程云琦
封面设计 魏　来 李　廉

冈特生态童书
小僵尸
[比]冈特·鲍利　著
[哥伦]凯瑟琳娜·巴赫　绘
朱　溪　译

记得要和身边的小朋友分享环保知识哦!
八喜冰淇淋祝你成为环保小使者!